NATIONAL GEOGRAPHIC

School Publishing

W9-BNS-023

PIONEER EDITION

By Rebecca L. Johnson

CONTENTS

Leapin' Lizards

By Rebecca L. Johnson

What's that leaping into the air?
Is it a furry mouse? A slimy frog?
No, it's a lizard with scaly skin.

as

Lizards are amazing animals. They can crawl, run, and climb. Some lizards can swim. And some can jump. Lizards come in many shapes, colors, and sizes.

Scientists **classify** living things. They put them into groups based on their characteristics, or **traits.** Lizards are **vertebrate** animals. Vertebrates have a backbone. Lizards are also **reptiles.** Snakes, turtles, and crocodiles are reptiles, too.

Scaly Skin

Different animals have different body coverings. Mammals have hair. Birds have feathers. Most amphibians have wet, slippery skin.

Reptiles have dry, hard **scales.** Scales cover a lizard's body. Some scales are bumpy. Others are smooth.

Other Lizard Traits

Lizards have tails. Most have four legs and feet with five toes. Many lizards also have claws.

Lizards have ears that look like little holes. They have many sharp teeth.

Most lizards have eyelids. But geckos do not. A gecko's eyes are covered by clear scales.

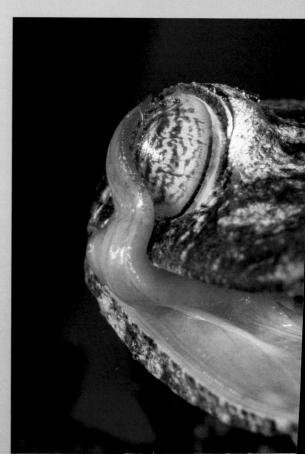

Quick Change Artist. This chameleon is a type of lizard that can change colors.

A Trick Tail

Many lizards can lose their tails. The tail breaks off if a predator grabs it. This gives the lizard a chance to escape. A new tail will grow back.

Fancy Feet

The basilisk lizard has skin between its toes. It can run on water. It can escape enemies by crossing streams or ponds.

The Rock Trick

The armadillo lizard is covered with thick, sharp scales. It can curl up into a spiky ball. Predators don't eat it because it looks like a sharp rock.

Twin Traits

Lizards are different from other animals. But lizards and other animals are alike in some ways, too. They can have similar adaptations.

Chameleons and monkeys have similar tails. Their tails can wrap around and hold on. That leaves feet free to do other things.

Sometimes it helps to be colorful. The anole lizard can show a flap of red skin. The frigate bird can puff out a red sac.

Looking bigger can scare predators. The frilled lizard and the cobra have a similar trick. They can make their heads look bigger.

No wings? No problem! A flying dragon lizard has special flaps of skin. So does a flying squirrel. They spread the flaps to glide through the air.

Leapin' Lizards

Show what you've learned about lizards by answering the questions below.

1 What other animals besides lizards are vertebrates?

2 Name three traits that lizards have.

3 Choose two of the lizards shown on pages 6 and 7. How are they alike? How are they different?

4 What is an adaptation?

5 Give an example of a lizard adaptation.